Alfred Addo-Yobo

New Module for deriving Maxwell's Four Relations

GRIN Verlag

Bibliografische Information der Deutschen Nationalbibliothek:

Die Deutsche Bibliothek verzeichnet diese Publikation in der Deutschen National-bibliografie; detaillierte bibliografische Daten sind im Internet über http://dnb.d-nb.de/ abrufbar.

Imprint:

Copyright © 2013 GRIN Verlag GmbH
Druck und Bindung: Books on Demand GmbH, Norderstedt Germany
ISBN: 978-3-656-57505-4

This book at GRIN:

http://www.grin.com/en/e-book/264225/new-module-for-deriving-maxwell-s-four-relations

GRIN - Your knowledge has value

Der GRIN Verlag publiziert seit 1998 wissenschaftliche Arbeiten von Studenten, Hochschullehrern und anderen Akademikern als eBook und gedrucktes Buch. Die Verlagswebsite www.grin.com ist die ideale Plattform zur Veröffentlichung von Hausarbeiten, Abschlussarbeiten, wissenschaftlichen Aufsätzen, Dissertationen und Fachbüchern.

Visit us on the internet:

http://www.grin.com/

http://www.facebook.com/grincom

http://www.twitter.com/grin_com

Derivation of the Four Maxwell's Relations

Alfred Addo-Yobo © 2013
Kwame Nkrumah University of Science & Technology

Abstract:

The four Maxwell's relations are important equations employed mainly in the field of chemical engineering to perform certain computations involving the four thermodynamic potentials, temperature (T), pressure (P), volume (V) and entropy (S)

In chemical engineering, the method for deriving these four relations is by employing the Gibbs-Duhem-Margules approach which is somewhat tedious and lengthy.

In this paper, we shall explore another module in the derivation of these four Maxwell's relations by employing certain simple techniques with our basis as the mathematical equation of the first law of thermodynamics.

Keywords: Maxwell's relations, 1st law of thermodynamics, chemical engineering, Gibbs-Duhem-Margules equations, derivation.

Introduction:

The basis for our method of derivation of Maxwell's relations is the mathematical equation for the 1st law of thermodynamics which is $\Delta U = \Delta q - \Delta w$…….. (1), where ΔU is the change in internal energy, Δq is the change in heat content and Δw is the change in the work done.

To express the 1st law of thermodynamics in terms of the four thermodynamic potentials, T, P, V, S, Δq which is the change in the heat content is expressed as TdS where T is the absolute temperature and dS is the total change in entropy. Δw is also expressed as PdV where P is the pressure applied and dV is the total change in volume.

Inserting the expressions above into equation (1), we have dU = TdS – PdV…(4). This is the mathematical equation of the 1st law of thermodynamics expressed in terms of the four thermodynamic potentials. With our basis together with some techniques, we shall derive the four Maxwell's relations which are as follows;

$$\left(\frac{\partial T}{\partial V}\right)_S = -\left(\frac{\partial P}{\partial S}\right)_V \ldots\ldots \text{ (a) } \left(\frac{\partial S}{\partial P}\right)_T = -\left(\frac{\partial V}{\partial T}\right)_P \ldots\ldots \ldots \text{(b)}$$

$$\left(\frac{\partial T}{\partial P}\right)_S = \left(\frac{\partial V}{\partial S}\right)_P \ldots\ldots \text{ (c) } \left(\frac{\partial P}{\partial T}\right)_V = \left(\frac{\partial S}{\partial V}\right)_T \ldots\ldots\ldots \text{ (d)}$$

Alfred Addo-Yobo's derivation:

This method of deriving the four Maxwell's relations aside having a basis as the mathematical equation of the 1st law of thermodynamics, the 1st of incomplete Maxwell's relations is derived with which the other three incomplete Maxwell's relations shall derived.

The mathematical equation of the 1st law of thermodynamics $dU= TdS - PdV$ is rewritten by applying partial differentials to all the potentials present in the mathematical equation above. Thus we have, $\partial U = \partial T\partial S - \partial P\partial V$.......... (5).

Setting $\partial U = 0$ in equation (5), we have $\partial T\partial S = \partial P\partial S$........... (6). Equation (6) is very significant in deriving the 1st incomplete Maxwell's relation which is key to deriving the other three incomplete relations.

A closer look at equation (6) shows two thermodynamic potentials each on the right and left hand sides of the equation. In order to derive the 1st incomplete Maxwell's relations, one thermodynamic parameter on the right hand side or left hand side respectively divides another thermodynamic parameter on the left hand or right hand sides respectively of the equation and vice versa.

As seen from equation (6), we have $\partial T\partial S$ on the left hand side and $\partial P\partial V$ on the right hand side. One parameter on the left hand, e.g ∂T divides another on the right hand, e.g ∂P and then the other on the right hand which is ∂V divides the other on the left hand which is ∂S.

From the description above it is clear that, such divisions may not produce the same results. Thus, the 1st of incomplete Maxwell's relations will be different for many but due to the uniqueness and versatility of the techniques to be unveiled, the final result will be the same four Maxwell's relations shown above.

NB: *Divisions to derive the first of incomplete Maxwell's relations can be done in any form to equation (6) but parameters on the same side of the equation cannot divide themselves.*

By the divisions performed earlier, **our chosen first of incomplete Maxwell's relations is** $\left(\frac{\partial P}{\partial T}\right) = \left(\frac{\partial S}{\partial V}\right)$. This is our first incomplete Maxwell's relations from which we shall construct the other relations.

There are several other outcomes when the divisions are done which are;

$$\left(\frac{\partial T}{\partial P}\right) = \left(\frac{\partial V}{\partial S}\right), \left(\frac{\partial T}{\partial V}\right) = \left(\frac{\partial P}{\partial S}\right), \left(\frac{\partial V}{\partial T}\right) = \left(\frac{\partial S}{\partial P}\right), \left(\frac{\partial P}{\partial T}\right) = \left(\frac{\partial S}{\partial V}\right).$$

Any of these can be chosen as the 1st incomplete Maxwell's relations from which the other three incomplete relations shall be derived.

Interchange technique:

In this technique, we shall derive the other three incomplete Maxwell's relations by applying the interchange technique to only the 1st incomplete Maxwell's relation.

$$\left(\frac{\partial P}{\partial T}\right) \neq \left(\frac{\partial S}{\partial V}\right) \qquad \longrightarrow \qquad \left(\frac{\partial P}{\partial S}\right) = \left(\frac{\partial T}{\partial V}\right)$$

This means, switch the position of the parameters along that diagonal.

$$\left(\frac{\partial P}{\partial T}\right) \searrow \left(\frac{\partial S}{\partial V}\right) \qquad \longrightarrow \qquad \left(\frac{\partial V}{\partial T}\right) = \left(\frac{\partial S}{\partial P}\right)$$

This means, switch the position of the parameters along that diagonal.

$$\left(\frac{\partial P}{\partial T}\right) \Downarrow \Downarrow \left(\frac{\partial S}{\partial V}\right) \qquad \longrightarrow \qquad \left(\frac{\partial T}{\partial P}\right) = \left(\frac{\partial V}{\partial S}\right)$$

This means, switch the position of the parameters along that vertical.

As seen the 1st law of incomplete Maxwell's relations was singly employed to derive the other three incomplete Maxwell's relations. Thus, the four incomplete Maxwell's relations derived by the interchange technique are;

$$\left(\frac{\partial P}{\partial T}\right) = \left(\frac{\partial S}{\partial V}\right)\ldots\ldots\text{(a)} \qquad \left(\frac{\partial P}{\partial S}\right) = \left(\frac{\partial T}{\partial V}\right)\ldots\ldots\ldots\text{(b)} \qquad \left(\frac{\partial V}{\partial T}\right) = \left(\frac{\partial S}{\partial P}\right)\ldots\ldots\ldots\text{(c)}$$

$$\left(\frac{\partial T}{\partial P}\right) = \left(\frac{\partial V}{\partial S}\right)\ldots\ldots\text{(d)}.$$

From this stage, shall proceed to apply the subscripts and signs to these four incomplete Maxwell's relations to derive the complete four Maxwell's relations.

Sign application:

Certain portions of the final four Maxwell's relations have negative signs attached to them and in this part of the paper, we shall a rule which will enable as fix these negative signs to the relations necessary.

Addo-Yobo's rule:

When two diagonal lines are drawn across any of the four incomplete Maxwell's relations and arrowheads are fixed at either ends of these lines depending on the progression of the alphabets, a negative sign is applied to either side of the equation for which the arrowheads point in different directions else they remain positive.

This rule means that, when we take any of the four incomplete Maxwell's relations and draw two diagonal lines to link the parameters, we must place arrowheads on the both diagonal lines.

The position of the arrowheads depends on the progression of alphabets. When we say "progression of alphabets" we mean when we take a particular diagonal and note the two thermodynamic parameters at the ends of the diagonal, we fix the arrowheads at the end depending on how the alphabet's progressed or the order of the alphabets represented as parameters.

For example, if a diagonal line within an equation connects parameters, ∂G and ∂R, the arrowhead is placed close to ∂R as shown ($\partial R \longleftarrow \partial G$). This is done to show that the alphabets represented as parameter, R and G, progresses from G to R. That is to say, when reading the alphabets, we move from G to R thus the arrowhead at R.

Applying the concept of this rule to the four incomplete Maxwell's relations;

$$\begin{pmatrix}\frac{\partial P}{\partial T}\end{pmatrix} \times \begin{pmatrix}\frac{\partial S}{\partial V}\end{pmatrix} \longrightarrow \begin{pmatrix}\frac{\partial P}{\partial T}\end{pmatrix} = \begin{pmatrix}\frac{\partial S}{\partial V}\end{pmatrix}$$

The above shows two diagonals, one linking ∂P and ∂V and the other diagonal linking ∂S and ∂T. From the order of alphabets, since P comes before V, the arrowhead is fixed at ∂V to show the order of alphabets for the two parameters for the 1st diagonal. In the second diagonal that is linking ∂S and ∂T and from the order of alphabets, the arrowhead is fixed at ∂T to show that S comes before T. From the above arrowheads shown above and with the Addo-Yobo's rule in mind, since the arrowheads point in the same direction (downwards), both sides of the equation stays positive.

4

$$\left(\frac{\partial P}{\partial S}\right)\times\left(\frac{\partial T}{\partial V}\right) \longrightarrow -\left(\frac{\partial P}{\partial S}\right)=\left(\frac{\partial T}{\partial V}\right)$$

Since arrowheads as shown above at the ends of the diagonals point in different directions, we apply a negative sign to either side of the equation according to the Addo-Yobo rule.

$$\left(\frac{\partial V}{\partial T}\right)\times\left(\frac{\partial S}{\partial P}\right) \longrightarrow \left(\frac{\partial V}{\partial T}\right)=-\left(\frac{\partial S}{\partial P}\right)$$

Since the arrowheads as shown above at the ends of the diagonals point in different directions, we apply a negative sign to either side of the equation according to the Addo-Yobo rule.

$$\left(\frac{\partial T}{\partial P}\right)\times\left(\frac{\partial V}{\partial S}\right) \longrightarrow \left(\frac{\partial T}{\partial P}\right)=\left(\frac{\partial V}{\partial S}\right)$$

Since the arrowheads as shown above at the ends of the diagonals point in the same direction, both sides of the equation stay positive according to the Addo-Yobo rule.

From the step above, that is applying the Addo-Yobo rule we have the following four incomplete Maxwell's relations;

$$\left(\frac{\partial P}{\partial T}\right)=\left(\frac{\partial S}{\partial V}\right)\ldots\ldots\ldots\text{(a)}$$

$$-\left(\frac{\partial P}{\partial S}\right)=\left(\frac{\partial T}{\partial V}\right)\ldots\ldots\ldots\text{(b)}$$

$$-\left(\frac{\partial V}{\partial T}\right)=\left(\frac{\partial S}{\partial P}\right)\ldots\ldots\ldots\text{(c)}$$

$$\left(\frac{\partial T}{\partial P}\right)=\left(\frac{\partial V}{\partial S}\right)\ldots\ldots\ldots\text{(d)}$$

The final stage of this method of deriving Maxwell's four relations is to apply the subscripts to each of the four incomplete Maxwell's relations above.

Application of subscripts:

From the four incomplete Maxwell's relations above, it can be seen that each has two brackets containing two thermodynamic potentials. The subscript of each bracket in a single relation is the denominator parameter of the other bracket in the same relation.

Applying this simple technique to the four incomplete Maxwell's relations concludes the derivation process.

$$\left(\frac{\partial P}{\partial T}\right)_V = \left(\frac{\partial S}{\partial V}\right)_T \quad -\left(\frac{\partial P}{\partial S}\right)_V = \left(\frac{\partial T}{\partial V}\right)_S \quad -\left(\frac{\partial V}{\partial T}\right)_P = \left(\frac{\partial S}{\partial P}\right)_T \quad \left(\frac{\partial T}{\partial P}\right)_S = \left(\frac{\partial V}{\partial S}\right)_P$$

This concludes that the final and complete four Maxwell's relations are;

$$\left(\frac{\partial P}{\partial T}\right)_V = \left(\frac{\partial S}{\partial V}\right)_T \quad \text{............ (a)}$$

$$-\left(\frac{\partial P}{\partial S}\right)_V = \left(\frac{\partial T}{\partial V}\right)_S \quad \text{............ (b)}$$

$$-\left(\frac{\partial V}{\partial T}\right)_P = \left(\frac{\partial S}{\partial P}\right)_T \quad \text{............ (c)}$$

$$\left(\frac{\partial T}{\partial P}\right)_S = \left(\frac{\partial V}{\partial S}\right)_P \quad \text{............ (d)}$$

Conclusion:

This method is specially designed with tools that can be related to easily such as alphabets, diagonal lines, arrowheads, etc. It's important to compare your results with the expected outcome which was earlier revealed.

As mentioned earlier, the first of incomplete Maxwell's relations may be different for many as there are four possible outcomes, however by applying the interchange technique, Addo-Yobo's rule and subscripts results in the same outcome of **four Maxwell's relations.**